WHY DO ANIMALS DO THAT?

WHY DO ANIMALS GROOM EACH OTHER?

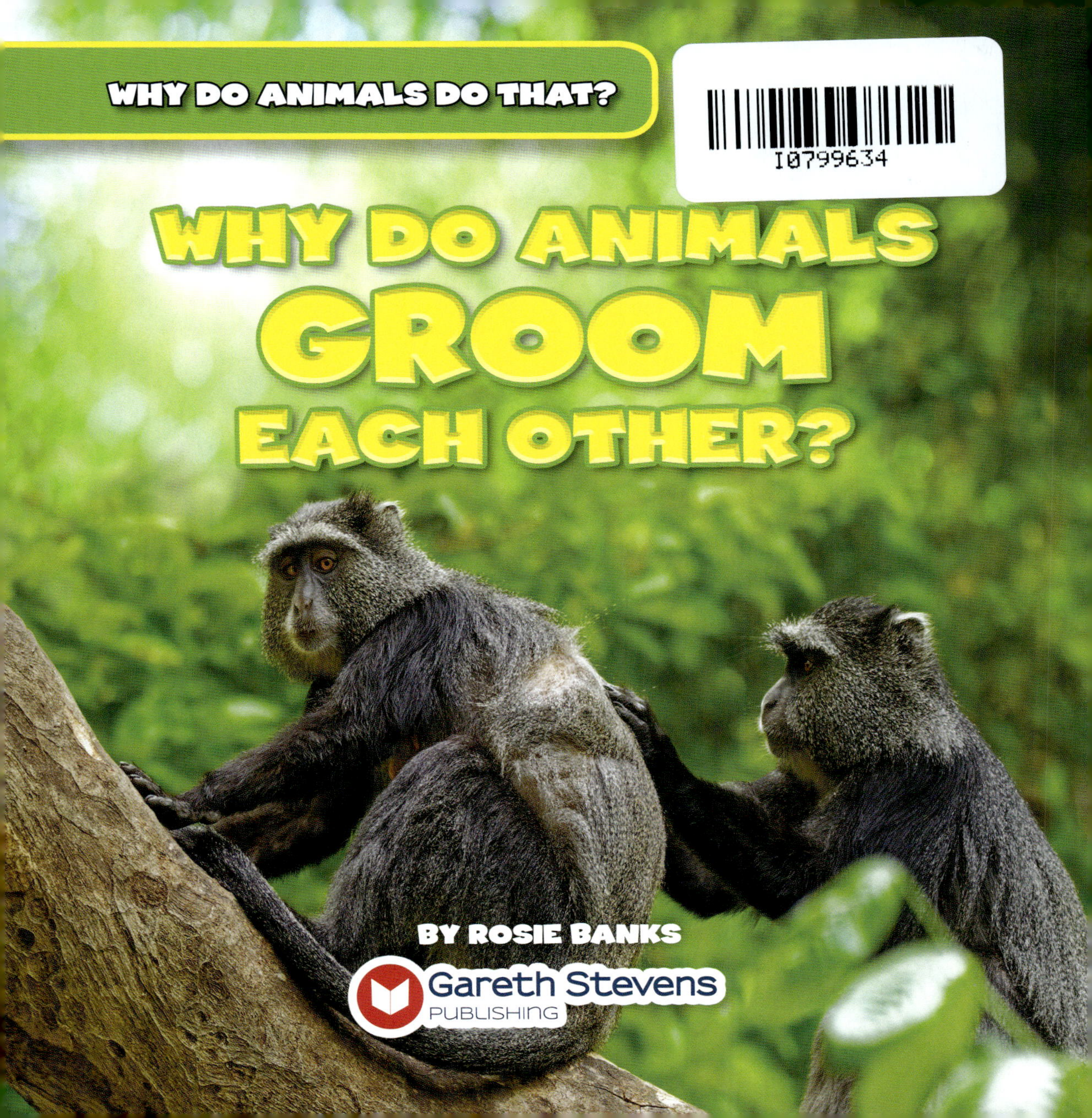

BY ROSIE BANKS

Gareth Stevens PUBLISHING

Please visit our website, www.garethstevens.com. For a free color catalog of all our high-quality books, call toll free 1-800-542-2595 or fax 1-877-542-2596.

Library of Congress Cataloging-in-Publication Data

Names: Banks, Rosie, 1978- author.
Title: Why do animals groom each other? / Rosie Banks.
Description: New York : Gareth Stevens Publishing, [2024] | Series: Why do animals do that? | Includes bibliographical references and index.
Identifiers: LCCN 2022045639 (print) | LCCN 2022045640 (ebook) | ISBN 9781538285121 (library binding) | ISBN 9781538285114 (paperback) | ISBN 9781538285138 (ebook)
Subjects: LCSH: Grooming behavior in animals–Juvenile literature. | Animal behavior–Juvenile literature.
Classification: LCC QL760 .B36 2023 (print) | LCC QL760 (ebook) | DDC 591.56/3–dc23/eng/20221003
LC record available at https://lccn.loc.gov/2022045639
LC ebook record available at https://lccn.loc.gov/2022045640

Published in 2024 by
Gareth Stevens Publishing
2544 Clinton Street
Buffalo, NY 14224

Designer: Claire Wrazin
Editor: Therese Shea

Photo credits: Cover Raphael Hagen/Shutterstock.com; Cover (background) W. Phokin/Shutterstock.com; Series Art (texture) Anna Timoshenko, p. 5 VGstockstudio/Shutterstock.com; p. 7 Wildnerdpix/Shutterstock.com; p. 9 jeep2499/Shutterstock.com; p. 11 Tomas Palsovik/Shutterstock.com; p. 13 Vladimir Turkenich/Shutterstock.com; p. 15 Anastasija Popova/Shutterstock.com; p. 17 010110010101101/Shutterstock.com; p. 19 Alicia Marvin/Shutterstock.com; p. 21 SomprasongWittayanupakorn/Shutterstock.com

Printed in the United States of America

CPSIA compliance information: Batch #CSGS24: For further information contact Gareth Stevens, New York, New York at 1-800-542-2595.

CONTENTS

Boldface words appear in the glossary.

Get Grooming

Has anyone ever told you to wash your face or fix your hair? They're telling you to groom yourself. Grooming means caring for your appearance. Animals also groom themselves and each other. Grooming is an important animal **adaptation**.

Primates Pick

Gorillas and monkeys are **primates**. Many primates live in groups. Primates in these groups use their teeth and fingers like combs. They pick out bugs and anything else they might find in each other's hair. Sometimes, they eat what they find!

GORILLAS

Grooming helps primates **bond** with each other. It makes them feel like they're part of their group. Grooming keeps primates healthy too. Bugs that are **parasites** can harm them. Plus, hair fluffed up from grooming can keep primates warmer!

MONKEYS

Lions Lick

You may have seen a cat licking its fur. Cats small and big groom themselves. Lions are the only big cats that live in groups. The groups are called prides. Lions in a pride groom each other. They lick and **nuzzle** one another.

A mother lion grooms her cubs. A male lion grooms his **mate**. Other lions in a pride groom each other too. This is a way for the pride to bond. Strong bonds are important. Lions protect, or guard, each other from other prides.

Horses Help

Horses also groom each other. They stand next to each other. Then they use their teeth to rub and softly bite. A study in Japan said wild horses groom each other to form strong bonds and get rid of parasites.

Birds Bond

Birds groom their feathers. Many birds have an oil **gland** near their tail. They use their beaks to spread oil from the gland on their feathers. The oil helps feathers bend. It makes them waterproof too. Bird grooming is called preening.

Some birds preen each other. This happens most often between mates. Preening may help mates **communicate** and bond. Preening can happen in bird flocks too. It may help the flock get along. Preening also gets rid of dirt and bugs.

Bugs Too!

Even some bugs groom each other. Some termites clean **fungus** off each other, for example. Removing the fungus from a few termites stops the fungus from spreading to the whole group. Grooming helps many animals in many ways!

HOW GROOMING HELPS

Removes bugs, dirt, and other harmful things from body

Helps mates bond

Fluffs up primate fur to keep them warm

termites

Helps animals in a group get along

Helps animals communicate

GLOSSARY

adaptation: A change in a type of animal that makes it better able to live in its surroundings.

bond: To form a relationship with someone or something. Also, the relationship that is formed.

communicate: To share ideas and feelings through sounds and motions.

fungus: A living thing that is somewhat like a plant, but doesn't make its own food, have leaves, or have a green color.

gland: A body part that produces something needed for a bodily function.

mate: One of two animals that come together to make babies.

nuzzle: To rub something gently with the nose or face.

parasite: A living thing that lives in, on, or with another living thing and often harms it.

primate: Any animal from the group that includes humans, apes, and monkeys.

FOR MORE INFORMATION

BOOKS

Andersen, Michelle Garcia. *Do Animals Have Families?* Vero Beach, FL: Rourke Educational Media, 2019.

Von der Gathen, Katharine. *Do Animals Fall in Love?* Wellington, NZ: Gecko Press, 2021.

WEBSITES

Preening: Why and How Birds Do It
www.thespruce.com/why-birds-preen-386448
Learn about preening and why it's important for a bird's health.

What Animals Preen?
animals.mom.com/animals-preen-3328.html
Find out more about animal grooming behaviors.

INDEX